SHIPS OF OUR NAVY
Carriers, Battleships, Destroyers, & Landing Craft

C. B. COLBY, ASSOCIATE, U. S. NAVAL INSTITUTE

Coward, McCann & Geoghegan   New York

ALL RIGHTS RESERVED. COPYRIGHT, 1953, BY C. B. COLBY

Printed in the United States of America

Fourteenth Impression

SBN: GB 698-30313-X

Foreword

In collecting and preparing the material and photographs for this book I was confronted with quite a problem. I wanted to show that the United States Navy was not merely a fleet of battleships, destroyers, torpedo boats, and minesweepers, but a fleet of many more types of surface craft than are usually thought of in connection with our fleets. I had no idea what I was getting into. I found that our Navy has well over 170 different designations of ships and boats, and that there is even an official difference between those two. I found that when a boat is designed too big to be carried upon the deck of another vessel then it automatically is a ship whether it likes it or not.

In selecting the 44 vessels for this book I have tried to pick the most important, interesting, or perhaps least known types, to help give you a new impression of just what our Navy is made up of as far as surface craft are concerned. Many of them may appear unexciting at first glance but I assure you that they are as important to our Navy's strength and collective punch-power as the often unimpressive "seconds" in the corner of a champion fighter. And that's just about what some of these vessels are.

They patch up the "champ" between rounds, keep him confident, coach him as to what to do in the next round, and generally keep him on his toes and off his back. There is plenty of glory in doing just that as the men and officers who man the "champs" they second will tell you. It's a great team from the top battleship down to the slowest and most unglamorous fleet auxiliary—a first-string team with *every* vessel manned by fighters.

As you go through this book I'm sure you will agree that there are some mighty interesting "seconds" backing up our fighters on the high seas, and that whether a fighter or the guy in the ring corner as a fleet auxiliary, they all rate the title of "Champ."

I would like to acknowledge with sincere thanks the kind co-operation of the following who helped me assemble the material and photographs for *Ships of Our Navy:* Mr. Orville S. Splitt, Chief, Book & Magazine Branch, Office of Public Information, Department of Defense; Lieutenant Theodore Taylor, USN, Head, Magazine & Book Branch, Office of Information, Department of the Navy; the U. S. Navy Film Library; and Lieutenant (JG) K. C. McCormick, USNR, Assistant Press and Photo Officer, Third Naval District.

C. B. COLBY,
Associate, U. S. Naval Institute

All illustrations are Official U. S. Navy Photographs

Contents

Foreword	2	High Speed Transport, USS *Begor*	26
New Super Carrier, USS *James Forrestal*	4	Amphibious Force Flagship, USS *Taconic*	27
Carrier, USS *Coral Sea*	5	Net Layer, USS *Abele*	28
Carrier, USS *Wasp*	6	Horn-Bowed Net Layer, USS *Etlah*	29
Carrier, USS *Antietam*	7	Wooden Minesweeper	30
Carrier, USS *Rendova*	8	Experimental Minesweeper	31
Battleship, USS *New Jersey*	9	Hospital Ship, USS *Rescue*	32
Heavy Cruiser, USS *Newport News*	10	Transport Barrack Ship (Towed)	33
Light Cruiser, USS *Worcester*	11	Ice Breaker, USS *Edisto*	34
Destroyer, USS *Forrest B. Royal*	12	Navy Fleet Tug, USS *Cherokee*	35
Tactical Minelayer, USS *Shannon*	13	Fleet Oiler, USS *Sabine*	36
Destroyer Escort, USS *Thaddeus Parker*	14	Fleet Tanker (Gasoline), USS *Nespelen*	37
Radar Picket, USS *Goodrich*	15	Repair Ship, USS *Ajax*	38
Submarine Chaser, PC-1263	16	Internal Combustion Engine Repair Ship, USS *Tutuila*	39
Landing Ship, Dock, USS *Fort Marion*	17	Battle Damage Repair Ship, USS *Midas*	40
Landing Ship, Medium	18	Ammunition Ship, USS *Paracutin*	41
Landing Ship, Medium (Rockets)	19	Surveying Ship, USS *Maury*	42
Landing Ship, Utility	20	Destroyer Tender, USS *Cascade*	43
Landing Ship, Vehicles, Personnel	21	Seaplane Tender (Large), USS *Currituck*	44
Patrol Torpedo Boat	22	Seaplane Tender (Small), USS *Valcour*	45
Blast Craft ("X-Craft")	23	Submarine Tender, USS *Orion*	46
Attack Cargo Ship, USS *Mathews*	24	Submarine Rescue Vessel, USS *Coucal*	47
Attack Troop Transport, USS *Bayfield*	25	How Our Navy Ships Are Named	48

Our Navy's Biggest

This artist's impression shows what our largest Naval vessel will look like when completed. The craft designated the CVA-58 ("CV" for Aircraft Carriers, and the "A" for Heavy Class) will be a 65,000-ton craft, 1,090 feet long and 236 feet wide, and the first completely flush-decked vessel to be built. Even the smokestacks will be horizontal along with such other items as radio antennas, etc. Speed will be about 33 knots. Some parts of the craft's superstructure are planned to be telescoping so that they may be raised or lowered as required for combat use or efficiency. Aircraft carriers are used to strike their offensive blows through bombing and fighting aircraft rather than guns and shells.

One of Our Toughest Customers

The great USS *Coral Sea* (CVB-43) is one of our biggest carriers. She and the USS *Midway* and USS *Franklin D. Roosevelt* are in a class by themselves. They displace (when fully loaded) over 55,000 tons, and are 968 feet long. The *Coral Sea's* hull has a "beam" (width from side to side) of 113 feet and an overall width of 136 feet. Besides her aircraft she is protected by a main battery of 14 five-inch guns plus a great many smaller caliber anti-aircraft weapons located about her hull and superstructure. She carries over 100 fighting aircraft, and her wartime crew numbers over 3,000 men and officers. A carrier's superstructure is called her "island."

Wasp with Many Stingers, a "Large" Carrier

The USS *Wasp*, designated CVB-18, is what is known as a short-hull Essex-class carrier. She is shown here with the planes and crew lined up "at quarters" for Captain's inspection. Forty-nine of her 100-plus fighting aircraft are shown in this lineup, and all but those required for operation of the ship, of her 2,500-plus crewmen. This great fighting floating airbase is 876 feet long, and can make 33 knots an hour. She is armed with a main battery of heavy guns plus a great quantity of smaller caliber rapid-firing anti-aircraft guns strategically located about her huge deck and hull. The aircraft of our carriers have four primary purposes regardless of the size of the carrier: to locate and observe enemy forces, carry out long-range attacks against an enemy, provide our own ships with protection from enemy aircraft, and spot submarines and bomb them.

Modified for Faster Takeoffs

The CVB-36 ("CV" for Aircraft Carrier, and the "B" for Large Class), the USS *Antietam*, has been recently modified to test a theory to speed up plane takeoffs. Aircraft must be on deck ready for takeoff ahead of time. They have formerly been crammed together down at one end of the deck, taking off from that end one at a time. This has meant a much shorter takeoff runway. With this "off set" runway, the waiting aircraft may move down the deck to this end along the righthand side while the ready aircraft are taking off at an angle, safe from the taxiing planes. This means that more aircraft may be on deck for takeoff than heretofore, and that aircraft have a longer runway all the time.

Small But Potent Pocket Carrier

This small but highly effective escort carrier type, the USS *Rendova*, was typical of many light carriers used during the Second World War. These light escort carriers were built up from merchant ship hulls and so could be turned out quickly and economically from a sort of "stock" hull design. As with all other carrier types, the CVE's ("CV" for Carrier and the "E" for Escort Type) usually required an escorting screen of cruisers and/or destroyers. Such CVE's as the USS *Rendova* are 557 feet long, and have an overall width of 105 feet. They carry about 30 aircraft and have a wartime crew of about 1,000 men and officers. Her main battery is small, usually 2 five-inch guns plus additional anti-aircraft weapons.

Super-Slugger of the Fleet

Our four biggest battleships (BB's), the *New Jersey* (shown here), the *Iowa*, the *Missouri*, and the *Wisconsin* are the heavyweights of the U. S. Navy. The duties of the BB's are to engage and sink any and all types of enemy ships by long-range gunfire, and to blast enemy shore and harbor installations by continuous and very heavy bombardment with their heavy guns. The USS *New Jersey* (BB-62) carries 9 sixteen-inch guns, 20 five-inch guns, 80 40 mm. guns and many .50 cal. machine guns as do others in her class. She is 887 feet long and has an extreme beam of 108 feet and she has a wartime crew of about 2,700 men and officers. Battleships can throw, in a single broadside of gunfire, more than ten tons of steel and explosives over 20 miles with great accuracy. Battleships are heavily armored above and below the waterline, and usually are screened by destroyers and light cruisers

The Heavy Cruiser Class

Cruisers are fast ships with long ranges but are lightly armored as compared with the battleships. They are classed as "CA" (Heavy), "CL" (Light), or as "CB" which stands for Large type. This type was started during World War II but did not get into full-scale combat tests. The CA-148 shown here is the USS *Newport News*, of the Heavy Cruiser class. She is 716 feet long, has an overall beam of 75 feet. She is armed with a main battery of 9 eight-inch guns (cruisers carrying guns of over 6 inches are classed as heavy cruisers) and 12 five-inch guns, plus 20 three-inch and other anti-aircraft weapons. She carries a wartime crew of over 1,850 men and officers and has a speed of about 32 knots. The speed and maneuverability of these boats make them very effective for sea operations and bombardment of shore installations. Her guns reach 14 miles or more.

Light Cruisers Are Shifty Fighters

One of the greatest advantages to the Light Cruiser as typified here by our speedy and long-range USS *Worcester* (CL-144) is her great range and high maneuverability. The range of light cruisers may be as much as 15,000 miles, which permits them suddenly to appear in unexpected places for bombardment of shore installations, participation in landing activities, or support of other missions, when least expected. The USS *Worcester* type is 679 feet long, 70 feet wide, and has a speed of over 32 knots. Her main battery is 12 six-inch guns, 20 three-inch guns and 12 40 mm. anti-aircraft cannon plus other light-caliber defense weapons. She carries a wartime crew of about 1,700 men and officers.

Wolf Pack of Our Fleet, the Destroyers

One of the most glamorous types of fighting ships has always been the destroyer, perhaps for its name, perhaps for its many exploits; but regardless, its reputation as a fast, shifty, and potent killer has been justly earned. There are several types of the "DD" (Destroyer) fighting ships, all basically similar. Here is the DD-872, the USS *Forrest B. Royal*, armed with a main battery of 6 five-inch guns and 5 twenty-one-inch torpedo tubes plus a great assortment of smaller weapons for anti-aircraft defense. Her length is 390 feet and her greatest width is 40 feet. Her speed is about 35 knots and her wartime crew is about 350 men and officers. Destroyers are used to form a protective screen about convoys, use their torpedoes against enemy craft, lay smoke screens, and act as scouts and landing-party support.

Tactical Minelayer Version of DD

Needing high-speed, highly maneuverable yet well-armed light minelayers, the Navy converted ten older model destroyers into what are termed tactical minelayers. These craft were fitted with tracks to handle wheeled mines, turntables to maneuver them from one track to another along the sides and stern of the craft, and a stern fitting to drop them overboard in enemy waters or in the path of an enemy fleet if need be. Each of these speedy well-armed DM's ("D" for Destroyer and "M" for Minelayer version) carriers 100 spherical mines plus usual DD-type armament—five torpedo tubes, 6 five-inch guns plus small-caliber anti-aircraft weapons. Note row of spherical mines alongside aft of rear stack. This is the USS *Shannon* (DM-25).

"Expendable" Destroyer, the DE

The "E" of the designation "DE" does not stand for Expendable as some might think, but for Escort. This fast sleek craft was designed during World War II as a quickly built escort vessel for anti-submarine patrol with convoys of vital merchant ships. They could be built for about one-third the cost of a regular destroyer, and so were turned out in great numbers for this important anti-sub assignment. They carried the most modern submarine detection and tracking devices, "hedgehogs" that hurled dozens of contact-firing depth charges where subs were detected, and a main battery of 2 five-inch guns plus assorted smaller-caliber weapons for defense. This is the DE-369, the USS *Thaddeus Parker*. She is 306 feet long with a beam of 36 feet, and carries a crew of 220 men and officers. Her top speed is about 24 knots.

Okinawa Campaign Invention, the DDR

The "R" stands for Radar Picket and this important class of destroyer-type vessel has become an essential part of the fleet. Equipped with early warning radar screens and equipment this speedy ship ranges far ahead of the main group to pick up attacking surface craft and aircraft of the enemy. It forms a sort of long-range advance guard that can warn the main fleet of approaching danger long before it reaches the larger group of ships behind it. The DDR-831, the USS *Goodrich*, carries a wartime crew of 350, and is armed with 6 five-inch guns plus a good supply of 40 mm. anti-aircraft weapons and .50 cal. machine guns. She was also fitted with a special tripod mainmast to carry the special radar equipment. She is 390 feet long, 40 feet wide, and has a speed of about 35 knots.

Submarine Chaser

Although some older types of "sub chasers" have been removed from the Navy list of active ships, the "PC" ("P" for Patrol and "C" for submarine Chaser) still is carried. This is particularly true of the 173-foot steel hull type as exemplified by the PC-1263. These long slender craft are diesel propelled, can make 18 knots, and are highly maneuverable. They are armed with a quick-firing, dual-purpose three-inch gun on the bow plus a quick-firing 40 mm. gun aft of the stack plus other smaller-caliber weapons. The "PC"-type craft usually stays close to harbors or works only with coastal convoys. They carry considerable detection and tracking devices to locate enemy underseas craft in the vicinity.

Seagoing Trojan Horse, the LSD

The largest of the LS (Landing Ship) types is the floating dry dock for delivering loaded landing craft into a beachhead area. Built around a basic floating dry dock, the LSD is loaded with landing craft within its hollow hull. When approaching the invasion area, the troops enter the landing craft, all equipped for combat. When at the proper place, the hold is flooded, the stern gate opens, and out sail the landing craft heading for the beach, loaded with troops ready for combat. The LSD's are more than 450 feet long and 72 feet wide. This is the USS *Fort Marion* (LSD-22). In her hold she can carry 14 LCM's, small troop-carrying craft a bit bigger than the LCVP on page 21. She is armed with a five-inch gun and a goodly number of 40 mm. twin anti-aircraft guns along her sides. She has a crew of about 250. Dock well extends forward under superstructure. Engines and crew space in dock walls.

Landing Ship and Landing Craft—the LSM

Although during World War II many types of landing craft were developed for specific missions and assignments, few of them are still active on the Navy list. Some of the original types have been improved. Here is the LSM ("L" for Landing, "S" for Ship, and "M" for Medium) fitted with the new Kirsten cycloidal propellers which permit it to turn completely about within its own length—a great advantage when maneuvering close to shores, rocky coastlines, etc. These LSM's carried troops, explosives, wounded, small tanks and vehicles, and practically anything required for a landing party. The bow lets down to form ramp to beach for men or vehicles. She is armed with anti-aircraft guns.

Rocket Tosser, the LSM(R)

This remarkable photo shows a cluster of twenty rockets hissing toward a distant target from the rocket launchers mounted aboard an LSM(R). (The "R" is for Rockets.) These LSM(R)'s were converted from standard LSM's by modifying the decks to accommodate as many as 105 rocket launchers, each of which could fire up to 30 five-inch rockets a minute. In addition to this terrific barrage of rocket-propelled explosives, these modified LSM's carried a dual-purpose five-inch gun and in some cases 4.2 mortars for further damage to shore installations. The five-inch gun could be used for offense as well as defense, and the mobility of these small vessels made them particularly effective as landing-party support craft.

Workhorse LSU

Some of our Navy's ships are converted into other-purpose craft by merely changing the designation. For example, these LSU's ("L" for Landing, "S" for Ship, and "U" for Utility) used to be designated LCT ("L" for Landing, "C" for Craft, and "T" for Tank) and were used to carry about six Army tanks on their open decks. They were not seagoing and were carried into action areas on LST's or in sections on transports. Normally, any landing vessels under 200 feet in length are classified as landing CRAFT, and those over are designated as landing SHIPS, but this small 120-foot landing craft is now rated as a landing ship and is used mainly for harbor and coastal cargo and other utility work. The device dangling over her stern is the anchor and supporting frame. She is armed with a small cannon. Bow forms ramp.

Testing a New LCVP

New materials as well as designs are constantly being developed and tested by the Navy's Bureau of Ships. Here, for example, a new design for a LCVP ("L" for Landing, "C" for Craft, "V" for Vehicle, and "P" for Personnel) is being tested in Puget Sound. In this case the "P" might very well stand for Plastic, for this new design is made of this material. The entire craft with the exception of the engine and some fixtures is made of plastic material; strong, impervious to salt water, and easily repaired. The bow of these small landing crafts can be lowered to form a ramp so that the troops carried may rush ashore as soon as the bow slides up onto the beach being stormed. This particular small craft is unarmed.

Pocket-Sized Power, the PT Boat

A new model for future PT boats gets a workout. The "P" is for Patrol and the "T" designates Torpedo. Those used so successfully during World War II were mostly of plywood with three Packard engines. Of over 800 manufactured, 70 were lost, 219 went to Russia, and the rest were sold or scrapped during 1946. The missions of the "PT" included laying smokescreens, aiding crashed aircraft at sea, transporting small groups of personnel such as commando parties, using their torpedoes against enemy surface craft under the cover of darkness or poor visibility, and generally making use of their high-speed and polo-pony maneuverability. They were armed with 40 mm. guns, machine guns usually of .50 caliber, and torpedoes. This model—811—is being tested for future development.

The Mysterious "X-Craft"

One of the many surprises prepared by our Navy for use against the enemy during World War II, but which never went into action, was the fantastic radio-controlled blast craft known as the X-Craft. "Operation Stinger" was the program aimed to clear enemy beaches and harbors of underwater obstructions by sinking explosive-laden ships to blow them up by radio. Some of these "drones" (obsolete cargo ships) controlled by radio from far at sea could carry as much as 7,000 tons of explosives that could blast underwater obstructions clear for following landing craft. One of the weird craft designed specifically for this type of work was this fantastic little "salamander," as it was also called. It could do better than 17 mph in water and over 15 mph on land. It carried 1,200 pounds of explosives and could be directed by radio to either sink and explode its load under water or climb up onto the shore and head for beach forts where it could be exploded to demolish them. The X-Craft could start, stop, speed up or slow down, turn around, and explode—all under radio control.

Attack Cargo Ship, the AKA

It is difficult to imagine a cargo ship attacking anything, but the AKA's did and with both fists. The "AK" is the desgination of a cargo ship and the second "A" signifies Attack. The USS Mathews (AKA-96) carries enough armament to enable her to unload her cargoes right along with other attacking vessels of all types. She carries a five-inch gun plus 8 40 mm. quick-firing anti-aircraft guns. These give her protection from enemy attack while she unloads her landing craft with her 30-ton booms so that they may form an attacking assault force. She goes right in close along with the other types of fighting ships and sets her assault forces over the side as fast as possible, while fighting off attacking forces with her own guns. AKA's average about 450 feet in length and are usually capable of speeds of over 16 knots.

APA's Are Attack Troop Transports

There is a whole series of "attack ships" of various purposes. The APA's are big, fast, heavily armed troop transports used to carry fully armed troops right into the combat area itself in the face of enemy action and firepower. The "AP" means Transport, and the second "A," as in the case of the attack cargo ship opposite, means Attack. APA's are armed with either a five-inch cannon or a three-inch gun plus a goodly assortment of smaller weapons for defense. They could carry about 1,500 fully equipped troops plus the crew of the ship itself, which might run to over 500 men. It also carried landing craft of various types. Notice the two LCVP's on aft deck of this APA, the USS *Bayfield* (APA-33). The rectangular-shaped objects hung about the sides of the *Bayfield* are life rafts ready for instant release in case of emergency.

Destroyer Into Troop Ship, the APD

One of the most interesting conversions of the Navy was the APD ("AP" for Troop Transport, and the "D," strangely enough, for Destroyer or High Speed, whichever you prefer), converted from either a destroyer or a destroyer escort for the speedy delivery of troops where needed in a hurry. You will note the similarity between the lines of this APD, the USS *Begor*, and the true destroyer escort on page 14. This APD-127 carries an interesting insignia by its number, a tricolored square. This means that this particular ship carries an underwater demolition team, or unit of "frog men" as they are popularly called. The high-speed transport is used to bring reinforcements to another part of the assault area if more men are needed in a hurry and they must be brought into battle under fire. Her dimensions are the same as for destroyer escorts, but armament is reduced to make room for landing craft carried amidships, unloading booms and other troop accommodations.

Amphibious Force Flagship, the AGC

The successful operation of any amphibious force depends upon the planning by its top officers, complete and reliable communications, and instant changes of plans and strategy to take advantage of changes in situations as they occur. To do this with efficiency a headquarters must be maintained right with the amphibious force. These Amphibious Force Flagships are known as AGC's and are especially designed for their assignment. Complete and well-equipped quarters for top officers are included and elaborate communications equipment for every type of radio contact are carried. A large number of speedy launches are carried on deck to ferry officers from ship to ship while conferences are under way. These AGC's are well armed with 2 five-inch guns plus an assortment of 40 mm. rapid-firing anti-aircraft weapons and other light arms. This AGC is the USS *Taconic* (AGC-17).

Layer of Nets, the AN

During war conditions many vital ports and harbors are protected from sneak submarine attacks by huge steel nets hung across the harbor entrance from big hollow metal floats. The laying of this type of net requires a special ship designed to carry and lay such a device. Of these there are several types and sizes. This layer, the AN-58, is the USS *Abele,* one of the larger types, capable of carrying the net across the ocean to where it is to be laid. These net layers are armed with a five-inch gun and other light weapons for defensive action if attacked, and are fitted with special devices and stowage room for the huge steel nets and the buoys to float them. The rectangular boxlike affairs held in the slanting frames on either side of the masts are quick-release life rafts.

Horn-Bowed Net Layer, an AN

The USS *Etlah*, AN-79 ("A" for Auxiliary, and "N" for Net layer), is one of the highly specialized Navy designs. These unusual-looking craft are designed to lay and tend the steel submarine nets placed about harbors and fleet anchorages. The net, attached to the big metal globes seen on the deck and in the water about the bow, is fed into the water between the "horns" protruding from the stubby bow of the craft. The boom of the crane is used to hoist the buoys aboard for servicing and making other repairs to the net. These nets come in sections and may be added to or made shorter as conditions require. This AN-type ship is usually unarmed, as it works in and about harbors or other armed craft at anchor.

Back to Wooden Ships, the AMS

There is one place in the Navy where wooden ships are far better than steel, and that is in the tricky and hazardous business of minesweeping. The AMS-368 is one of the all-wooden craft designed to be safe from setting off magnetic mines developed during World War II. These mines were exploded by magnetic reaction to the steel hull of a passing ship even without actually touching the mine itself. The AMS-368 ("A" for Auxiliary, and the "M" for Minesweeper with the added "S" for wooden-hulled—or perhaps "safe" if you prefer) cruises back and forth in waters believed to be mined, and by dragging long cables behind, it locates and brings to the surface these deadly affairs. The davits at the stern are for lowering the "kites" into the water—note white-slotted affair hanging from nearest davit. These devices when attached to the cables move out to either side of the sweeper under water, much in the manner of kites rising on the wind. It is these kite "strings" of steel cable that locate the mines.

Experimental AMS

Not much can be told about this new small minesweeper, with the exception that it is undergoing tests. The davits on the stern are used to raise and retrieve the cylindrical paravanes that carry the steel mine-locating cables out to either side of the craft. The paravanes are those white "blimp-shaped" affairs dangling from the davits on the stern. Another device is being taken aboard. The cables to these devices run over the big pulleys you can see; one in silhouette like a hanging black ball and the other on this side of the stern just over the device still in the water. The heavy cable on the huge reel amidships is let out astern to a distance of several hundred feet. A device on the end of it sends out impulses to set off any enemy acousticle mines in the vicinity, at a safe distance behind the minesweeper. The AMS-368 (opposite) also carries such a cable just aft of derrick.

No Mistaking the AH

Painted white, with a broad green stripe running clear around it plus huge red crosses, a hospital ship is easy to spot under any conditions. The "A," as before, stands for Auxiliary (of the fleet) and the "H" of course identifies it as a Hospital Ship. By international law and agreement such ships as the USS *Rescue* (AH-18) are not usually attacked—but they have been fired upon. They are always brightly lighted at night and never carry weapons or combat troops. They are as well equipped as a modern shore hospital, and can take care of many hundreds of sick or wounded. Besides the regular crew of line officers and men, there is a full staff of doctors, nurses, dentists, surgeons, chaplains, and specialists, plus numbers of orderlies and other medically trained enlisted men. A sister ship, the USS *Haven*, covered the Bikini tests. A few of them are in operation, but during peacetime the majority of these AH vessels are held in reserve.

APL Designates "Towed Hotel"

One of the most awkward-appearing and certainly an unusual-looking vessel is the Transport Barrack Ship. There are two types of these fleet auxiliaries—the APB, which is a self-propelled floating barracks, and the APL type, which is towed. In both cases the "A" is for Auxiliary, as with all fleet auxiliaries, the "P" is for Transport type and the "B" and "L" identify it as self-propelled or towed. The ship shown, the APL-11, carries several thousand troops plus a small crew. She is armed with four small-caliber cannon or heavy-caliber machine guns, mainly as anti-aircraft protection. The framework on her top deck is for fair-weather awnings. Such craft are not suited for heavy seas and as they must be towed from place to place are used mainly to augment shore barracks when more room is needed temporarily.

Floe Crusher at Work, the AGB-2

With our Navy operating from the tropics to the arctic you are bound to find such highly specialized craft as icebreakers among the more standard Naval craft. This is the USS *Edisto* (AGB-2) in the ice of the arctic. These icebreakers are designed with a slanting bow so that they can smash into an ice floe and ride up on top of it. The bow and bottom plates are of inch-and-a-half heavy steel to withstand the impact. Once up on the ice, water ballast is pumped forward to increase the weight of the ship so that the floe breaks through from the thousands of tons of weight. Then the ship backs off and makes another charge at the ice until a path is broken through for the rest of the ships to follow in open water. This ship carries a helicopter, seen just aft of stub mast to rear of stack, to aid in scouting easiest path through floes ahead.

Navy Tugs Are AT's

There are four classes of Navy tugs bearing the AT designation. They are the ATF's ("A" for Auxiliary, "T" for Tug, and "F" for Fleet), useful for combat operations as they have good fire-fighting and salvage equipment and a very long range. Then there are the ATA's, which are fairly new, with the second "A" standing for Auxiliary, meaning that while they are long range they do not have all the equipment of the ATF's. There is also the ATO, with the "O" standing for Old, for they go back to World War I and are not capable of long-range operations. The last type is the ATR or "ocean tug, rescue" ("R") which, while of comparatively short range, is loaded with fire-fighting and salvage equipment. The ship shown here is the USS *Cherokee*, the ATF-66. The tripod mast and long boom aft of the stack are part of her salvage gear.

"Blood Bank" of the Fleet, the Oilers, AO's

Without the constant services of the gasoline and oil tankers, the fleet would be greatly limited in range and operations. These vital craft must keep up with all fleet operations and see that the fuel tanks of every ship within range are full to capacity and kept that way as far as possible. This is the USS *Sabine* (AO-25). The "O" stands for Oiler although most oil tankers of the fleet carry gasoline in varying amounts, to be able to refuel aircraft or vehicles using this fuel instead of the diesel oil. The AO's are usually faster than the AOG ("G" for Gasoline) tankers. The AO's make almost 20 knots, while the slower gasoline carriers average about 10 to 15 knots. The AO's are usually heavily armed against attack, for they are obviously a prime target for enemy aircraft and submarines. These vessels are equipped to refuel ships at sea while underway and without stopping. This is particularly essential when they are operating in enemy waters.

The Gasoline Tanker Is an AOG

With the letter "G" added to the designation of the fleet oil tanker (AO), shown on the opposite page, the gasoline tanker is labeled as to assignment. The USS *Nespelen*, shown off the coast of Greenland, carries high-test gasoline, lower-grade gasoline, and some oil. She is used mainly to refuel aircraft carriers and refuel shoreside installations where gasoline is a prime fuel. She can carry a crew of 124 and is armed with 4 three-inch guns and some machine guns for anti-aircraft use. Most AO's are heavily armed, usually with a five-inch gun, 4 three-inch cannon as well as at least 4 twin 40 mm. rapid-firing anti-aircraft guns. Due to complicated compartmentation and the latest in fire-fighting equipment, service aboard these AO's and AOG's is not so hazardous as you might think.

The Repair Ships (AR's) Go Right Along

As long as damaged vessels cannot make it back to a repair center, the next best thing is to take the repair center right along with the fleet. The "AR" series of vessels does just this, with every member designed to make special repairs. There is the ARH, for making hull repairs; the ARB, which patches up battle damage; the ARG, which specializes in landing craft repairs; and the ARV, which handles damaged aircraft from the carriers and other ships. All Navy vessels are equipped to make much of their own repairs of course, but the AR ships take over what the individual ships cannot handle. This is the USS *Ajax* (AR-6). She is about 530 feet long and has a beam of 73 feet. She carries a crew of nearly 1,300 men and is armed with 4 five-inch guns in bow and stern turrets. She carries heavy-duty cranes and is equipped to tackle about any type of repairs. Some of these AR's even carry forges and foundries for fabricating and casting heavy parts to replace those damaged in battle.

First-Aid Station for Engines, the ARG

With such a great proportion of all Navy equipment mechanized, it is no wonder that a special Internal Combustion Engine Repair Ship had to be developed to handle repairs of such power plants. This is the USS *Tutuila* (ARG-4) underway. The ARG's can repair or rebuild every type of gasoline and oil engine used by landing craft, landing ships, winches, launches, and even bigger types of power plants. The modern AR is armed with a usual five-inch gun on the stern, plus several 40 mm. anti-aircraft weapons located about her decks. They carry many types of repair technicians, including optical and electrical experts of various types, and expert carpenters and woodworkers along with the metal workers. The AR's keep our fleet ships in top repair and operating condition.

The ARB Works Under Fire If Need Be

Battle damage comes to almost every Navy combat ship sooner or later, and the next step is contact with the nearest shipyard, repair depot, or ARB. The ARB ("B" for Battle Damage) was designed about the hull of an LST (Landing Ship, Tank) and is able to run right into battle areas to make emergency repairs. These ex-LST's are manned by experts in quick effective repairs of all types, and carry machine shop equipment, welding equipment, and supplies to make all types of "first aid" repairs to keep the damaged vessel in operational condition. This is the USS *Midas* (ARB-5), camouflaged for service in enemy waters. She is armed, as were most original LST's, with three-inch guns and 40 mm. antiaircraft weapons. The ARB-5 has bow-opening doors as do all LST's, carries a crew of about 300, and is diesel powered. She is not a fast ship; in fact the LST's were often referred to as "Large Slow Targets" by the men who manned them in enemy waters. Later models of this ARB will undoubtedly be faster, more heavily armed, and generally even more effective.

The AE's Are Floating Volcanoes

This is literally the truth, for with the exception of one of the Ammunition Ships (the AE-20), all are named after active volcanoes. The "E" stands for Ammunition ship, or Explosives (if you find it easier to remember), and of the 16 in service at one time, almost all of them are now in reserve. These vessels were rather heavily armed, for they were considered a prime target by an enemy. Their armament included a five-inch gun aft and several smaller-caliber weapons for anti-aircraft use scattered about their decks. This is the AE-18, the *Paricutin*. Her twelve booms enabled her to load and transfer her lethal cargo with great speed. She could carry about 5,000 tons of explosives.

The "S" of AGS Stands for Surveying

One of the most important phases of fleet operations is the knowledge of where it can go in safety, anchor in safety, and maneuver without danger of running aground or hitting an underwater obstruction. Home waters are familiar of course and excellent charts are available for navigation. However, the Navy is by no means limited in operation to familiar waters, hence the development of the AGS series. These Surveying Ships are fully equipped to go into strange waters and make a complete detailed survey of navigating conditions, currents, reefs, harbors, bays, and inlets, and anything else that might be of importance to other ships of the fleet. On deck they carry various types of motor launches, both open cockpit and cabin types, and even amphibious craft to permit close surveys of coastlines. The USS *Maury*, shown here, is only lightly armed with small-caliber weapons as her work is mostly peacetime. Her aft deck may be used as a helicopter landing area.

Destroyer Tender, the AD

Even the rough and tough destroyers require a bit of "mothering" from time to time, in the way of repairs, servicing, and replacement parts for equipment or weapons. The AD's or Destroyer Tenders take care of this chore. They are equipped to supply all the items and services that the destroyers under their care cannot do for themselves. This includes repair machinery, spare parts, medical and dental services, chaplain service, and a host of other items and services a destroyer cannot carry or replenish itself. The destroyer may either tie up alongside the tender, or small boats may travel back and forth ferrying the needed items. This is the USS *Cascade* (AD-16) and she is armed with 4 five-inch guns and several smaller weapons for protection from enemy aircraft. Her heavy-duty cranes and many small boats carried upon her decks are used in her work.

AV's Are Seaplane Tenders

The big flying boats of the Navy occasionally need supplies at sea, far from shore bases where they can be serviced. It is then that they rendezvous with the AV servicing their unit. This is the USS *Currituck* (AV-7) alongside the ice pack in the antarctic. The *Currituck* is here lowering a giant Martin PBM Mariner flying boat into the water after servicing. The AV-7 is 540 feet long with a beam of 69 feet. She has a helicopter landing platform on her forward deck and complicated radar and electronic gear to bring long-range aircraft to her safely from long distances. She is armed with 4 five-inch guns and smaller weapons against enemy aircraft. She carries a crew of over 1,200 men and has a speed of over 19 knots. One of her sister AV's, the AV-11, the USS *Norton Sound*, has been converted into a guided-missile launcher.

Junior-Sized AV, the AVP

Although the "P" does not stand for Pocket Size, the small aircraft tender is much smaller than the regular AV's. This is the USS *Valcour* (AVP-55) and she is 310 feet long with a beam of 41 feet. These smaller AV's can get to downed aircraft with emergency supplies, fuel, and minor repairs much quicker than can the regular AV's, not because of speed but because there are over twice as many of them strategically located where they can be dispatched to aircraft in need of servicing. The *Valcour* carries a crew of about 215 and is armed with 2 three-inch guns and smaller weapons as anti-aircraft defense. She carries special radio equipment for making contact with aircraft and directing them to her from a distance, and she also carries small boats especially designed for use about aircraft upon the water.

Sub-Pack Mother Ship, the AS

In order to keep submarines in operational condition at all times, even if they cannot return to a home base for repairs, new parts, fuel, and attention to the crew in case of injury, the Submarine Tender was developed and given the designation of AS; the "A" as usual for fleet Auxiliary and the "S" to designate it as a Submarine Tender. The USS *Orion* (AS-18) is 530 feet long and has a beam of 73 feet. She carries a crew of about 1,300 men and officers and is armed with 4 five-inch guns in turrets fore and aft. She also has smaller-caliber weapons for protection from enemy aircraft. Her heavy-duty crane aft and many small boats carried on deck are part of her servicing equipment. Submarines may tie up alongside, several at a time, for attention.

Submarine Savers Are ASR's

Looking very much like the Fleet Tug (page 35), the Submarine Rescue Vessel bears the designation of ASR: "A" for Auxiliary, "S" for Submarine, and the "R" for Rescue. This is the USS Coucal (ASR-8). She is 251 feet long with a beam of 42 feet. She carries a crew of about 102 and is armed with 2 three-inch weapons plus smaller calibers. She is equipped with the latest in submarine rescue and salvage equipment including diving bells, powerful hoists, compressed air hoses and pumps to blow out flooded submarine compartments, trained divers, and rescue technicians. The ASR's are a great morale booster for the wearers of the Golden Dolphins—the submariners.

How Our Navy Ships Are Named

Battleships (BB) named for states.

Cruisers (CA,CL) named for cities and towns of the United States and capitals of United States' territories and possessions.

Cruisers (CB) named for territories and insular possessions of the United States.

Aircraft Carriers (CVB, CV, CVL) with few exceptions named for famous former ships of the Navy or important battles.

Aircraft Carriers (CVE) named for islands, bays, and sounds of the United States or for battles of World War II.

Destroyers (DD) named for deceased personnel of the following classes: American Navy, Marine Corps, and Coast Guard. Also named for personnel who have rendered distinguished service to the country, Secretaries and Assistant Secretaries of the Navy, and members of Congress closely identified with affairs of the Navy, and also famous inventors.

Destroyer Escorts (DE) named in honor of members of the Navy, Marine Corps, and Coast Guard killed in action in World War II.

Submarines (SS) named for various fish and other creatures of the sea.

Hospital Ships (AH) named with words suggestive of their missions—*Sanctuary, Haven, Consolation,* etc.

Cargo Vessels (AK) named for stars and planets, and counties of the various states.

Repair Ships (AR) named for mythological characters.

Ammunition Ships (AE) usually named for volcanoes.

Naval Transports (AP, APA, APR) named in honor of a great variety of people: deceased Commandants of the Marine Corps and Marine Corps officers; signers of the Declaration of Independence; famed women of history; and famous foreign men who rendered aid to our country in her earlier years. Also for counties in the United States and spots of historical interest.

Submarine Tenders (AS) named in honor of early pioneers in the development of the submarine and also for mythological sea deities.

Minesweepers (AM) named for various birds and with words descriptive of their activity and maneuverability, such as *Scurry, Velocity,* and *Skirmish.*

Submarine Rescue Vessels (ASR) also named for birds.

Gunboats (PG) named for small cities and words similar to those used for Minesweepers.

Oilers (AO) named for rivers with Indian names.

Seaplane Tenders (AV) named for various Sounds along our coasts.

Seaplane Tenders (AVP) named for bays, straits, and inlets.

Net Tenders (YN) named for various trees and in some instances, noted Indians.

Destroyer Tenders (AD) named for areas and various localities of our country, such as *Klondike, Yellowstone, Prairie,* etc.

Oceangoing Tugs (ATF) named for Indian tribes.

Harbor Tugs (YTB) named for Indian Chiefs and words of various Indian dialects.

Frigates (PF) named for cities and towns of our country and its possessions.